HOW'S THAT HUMAN?

Cover Design: Rita Claire and AM Conroy
Opening Page: Rita Claire and AM Conroy
Illustrations: AM Conroy

Copyright © 2021 HTH Publishing Inc.

All rights reserved. No part of this publication may be reproduced, stored in a retrieval system, or transmitted, in any form or by any means, electronic, mechanical, photocopying, recording, or otherwise, without prior written permission from the publisher. No part of this book may be reproduced in any manner whatsoever without written permission.

How's That Human? Chemistry - 1st Edition (softcover)

ISBN 9798581026199

Published by HTH Publishing Inc.
www.howsthathuman.com

THANK YOU!

Mom and Dad
I want to thank you for always believing in me and supporting my goals in life no matter what they were or where they took me! I love you!

Friends and Family
Thank you for your continued enthusiasm for my ideas! Thank you for not only pushing me but helping me grow, test, and develop this book!

My Team
I couldn't have asked for a better team to make this dream a reality! Your hard work was infectious, teamwork makes the dream work!

Former Students
I can't thank you enough honestly! You pushed me to be better and I am forever grateful for your influence in the teacher I am today!

YOU!
Thank you for taking a chance on a science book that looks at the human side of things and why science is ... well... YOU!

CHEMISTRY IS...

Introduction and History
Pg. 1-2

Tools and Safety
Pg. 3-5

Part 1: Atoms, Elements, and Matter
Pg. 6-10

Part 2: Compounds and Bonding
Pg. 11-14

Part 3: States of Matter and Chemical Reactions
Pg. 15-19

Part 4: Mixtures and Nutrition
Pg. 20-23

Part 5: Body Chemistry
Pg. 24-35

...YOU!

Appendices
Pg. 36-41

INTRODUCTION

To understand what it's like to be human, you need to understand the science of chemistry. Everything you are and do is based on the chemical interactions of atoms!

QUICK QUESTION: What is chemistry?

→ The study of matter: what it is made of, its properties, and how it changes.

Here are some reasons chemistry is important to you:

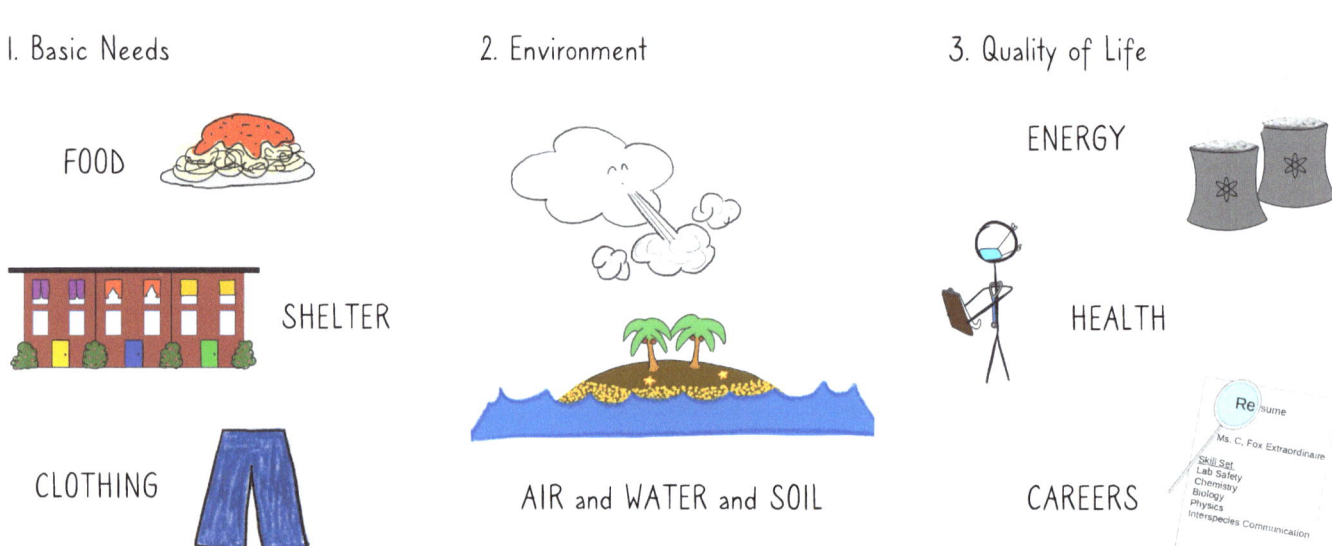

1. Basic Needs
   - FOOD
   - SHELTER
   - CLOTHING

2. Environment
   - AIR and WATER and SOIL

3. Quality of Life
   - ENERGY
   - HEALTH
   - CAREERS

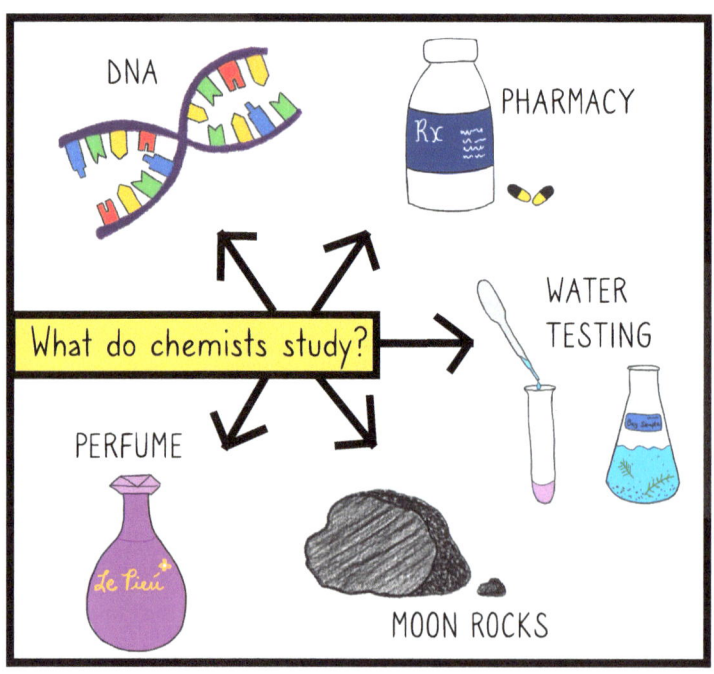

What do chemists study?
- DNA
- PHARMACY
- WATER TESTING
- MOON ROCKS
- PERFUME

Scientist Spotlight

Dr. Segenet Kelemu is an Ethiopian scientist who works as a molecular plant pathologist. She uses chemistry and life science to research ways to improve agriculture and crops. Coming from humble beginnings in a small village, she was the first woman from her region to go to college! She is an inspiration for all!

HISTORICAL TIMELINE

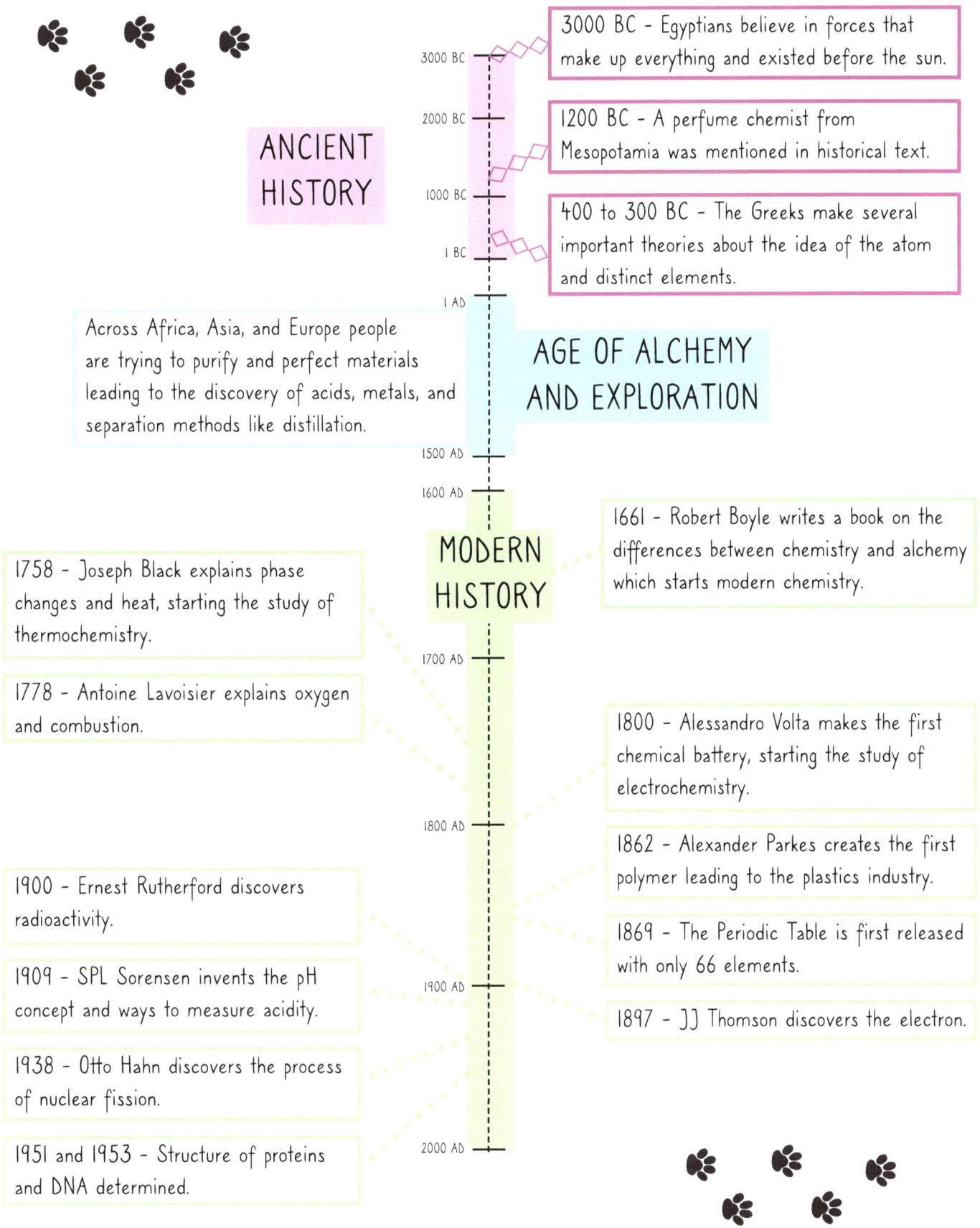

ANCIENT HISTORY

3000 BC - Egyptians believe in forces that make up everything and existed before the sun.

1200 BC - A perfume chemist from Mesopotamia was mentioned in historical text.

400 to 300 BC - The Greeks make several important theories about the idea of the atom and distinct elements.

AGE OF ALCHEMY AND EXPLORATION

Across Africa, Asia, and Europe people are trying to purify and perfect materials leading to the discovery of acids, metals, and separation methods like distillation.

MODERN HISTORY

1661 - Robert Boyle writes a book on the differences between chemistry and alchemy which starts modern chemistry.

1758 - Joseph Black explains phase changes and heat, starting the study of thermochemistry.

1778 - Antoine Lavoisier explains oxygen and combustion.

1800 - Alessandro Volta makes the first chemical battery, starting the study of electrochemistry.

1862 - Alexander Parkes creates the first polymer leading to the plastics industry.

1869 - The Periodic Table is first released with only 66 elements.

1897 - JJ Thomson discovers the electron.

1900 - Ernest Rutherford discovers radioactivity.

1909 - SPL Sorensen invents the pH concept and ways to measure acidity.

1938 - Otto Hahn discovers the process of nuclear fission.

1951 and 1953 - Structure of proteins and DNA determined.

CHEMISTRY TOOLS

To measure volume chemists use:

  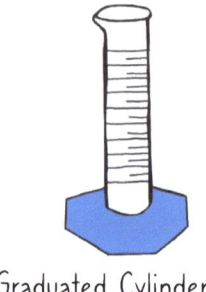  

Beakers   Flasks   Graduated Cylinders   Rubber Stoppers (as lids)

 **QUICK QUESTION:** What is volume?

->The amount of space something takes up. Chemists usually measure fluid in milliliters, or mL, that are listed on the side of the glassware.

Some other important glassware:

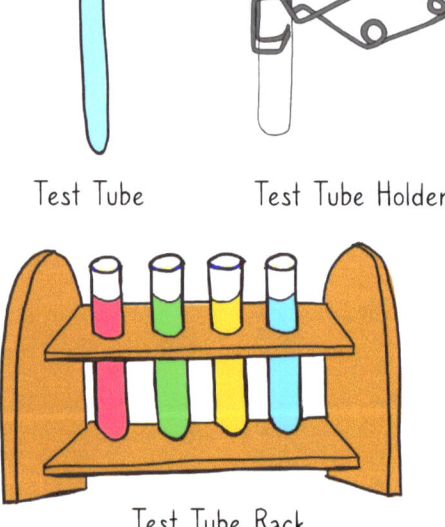

Test Tube   Test Tube Holder

Test Tube Rack

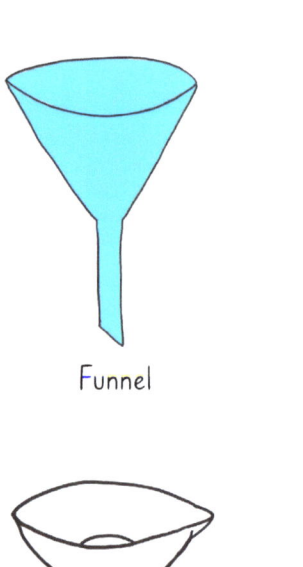 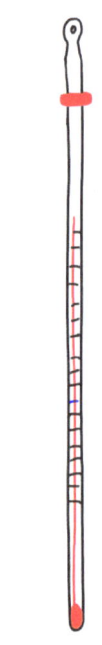

Funnel

Evaporating Dish   Thermometer

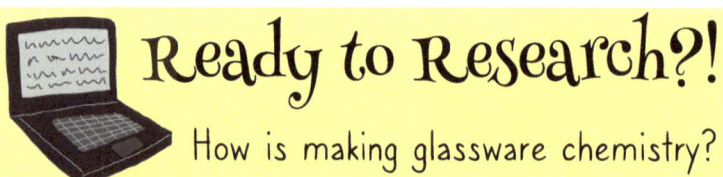

Ready to Research?!

How is making glassware chemistry?

Search Suggestions: scientific glass blowing, chemistry of colored glass

DID YOU KNOW: Glass blowing has been traced back to the Roman Empire?!

CHEMISTRY TOOLS

Hand tools include:

Tongs   Mortar and Pestle   Dropper   Pipette   Wash Bottle   Spatula

To measure weight or mass:

Electronic Scale

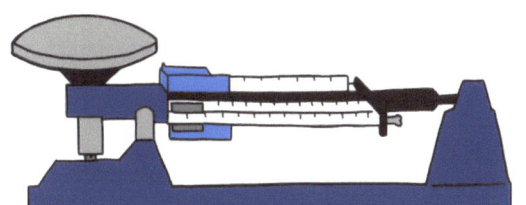

Triple Beam Balance

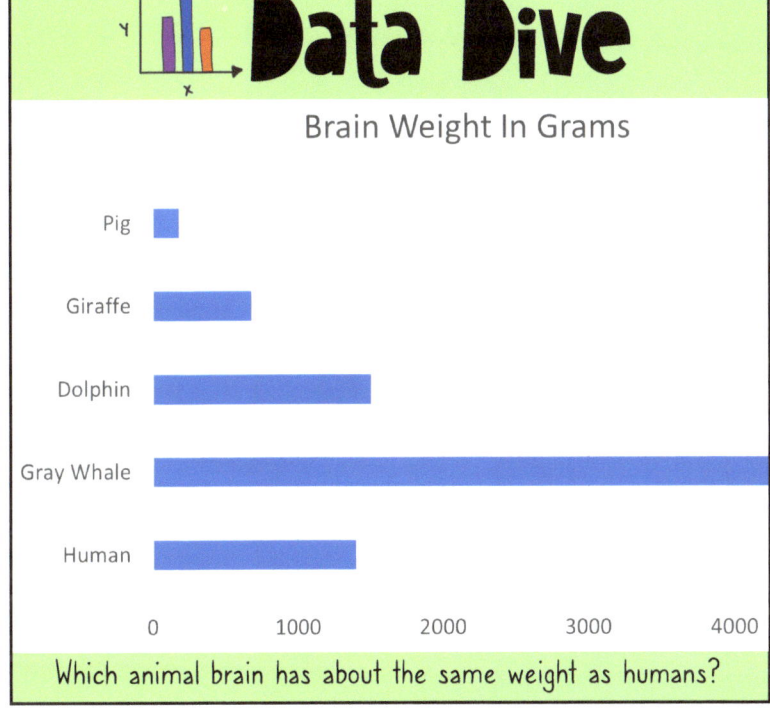

Data Dive

Brain Weight In Grams

(Bar chart showing: Pig, Giraffe, Dolphin, Gray Whale, Human; x-axis 0 to 4000)

Which animal brain has about the same weight as humans?

To study many reactions at once:

24 Well Plate

Some tools work together:

A. Ring Stand
B. Ring Clamps
C. Wire Gauze
D. Clay Triangle
E. Bunsen Burner

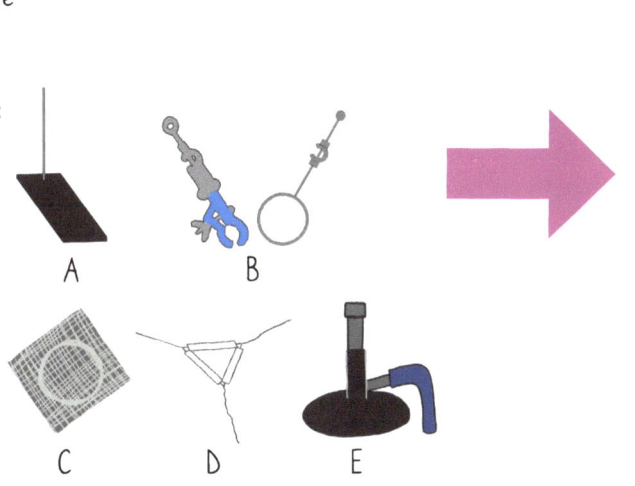

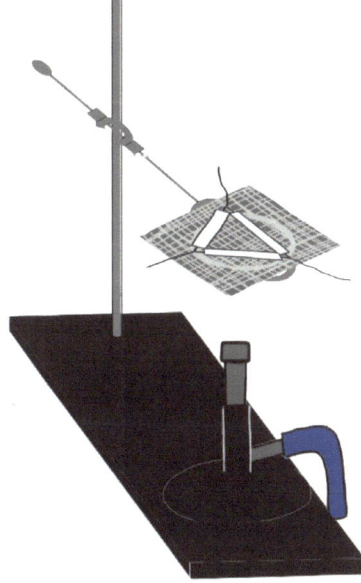

SAFETY AND SCIENCE

Wear long sleeves, like a lab coat!

Wear gloves!

Wear pants!

Wear closed-toe shoes!

Pull back long hair!

Don't forget eye protection!

PART 1

ATOMS & ELEMENTS MATTER

"Hi I'm the element oxygen!"

"And I'm the element magnesium!"

"Together we make the compound magnesium oxide, or MgO."

Chemical Properties are the result of a chemical reaction changing the compound. For example, MgO can be used as an antacid, as an insulator, or to make plastics.

Physical Properties are things we can measure or use our senses to describe. For example, MgO is a white solid powder, is made up of crystals, and has a mass of 40 grams.

 **QUICK QUESTION:** What is matter?

→ Anything that takes up space and has mass.

"Say it out loud with us!!!"

SECTION 1: The Atom

MATTER is made up of atoms, or atoms are the building blocks of MATTER. They take up space, have mass, and make up everything in the universe!

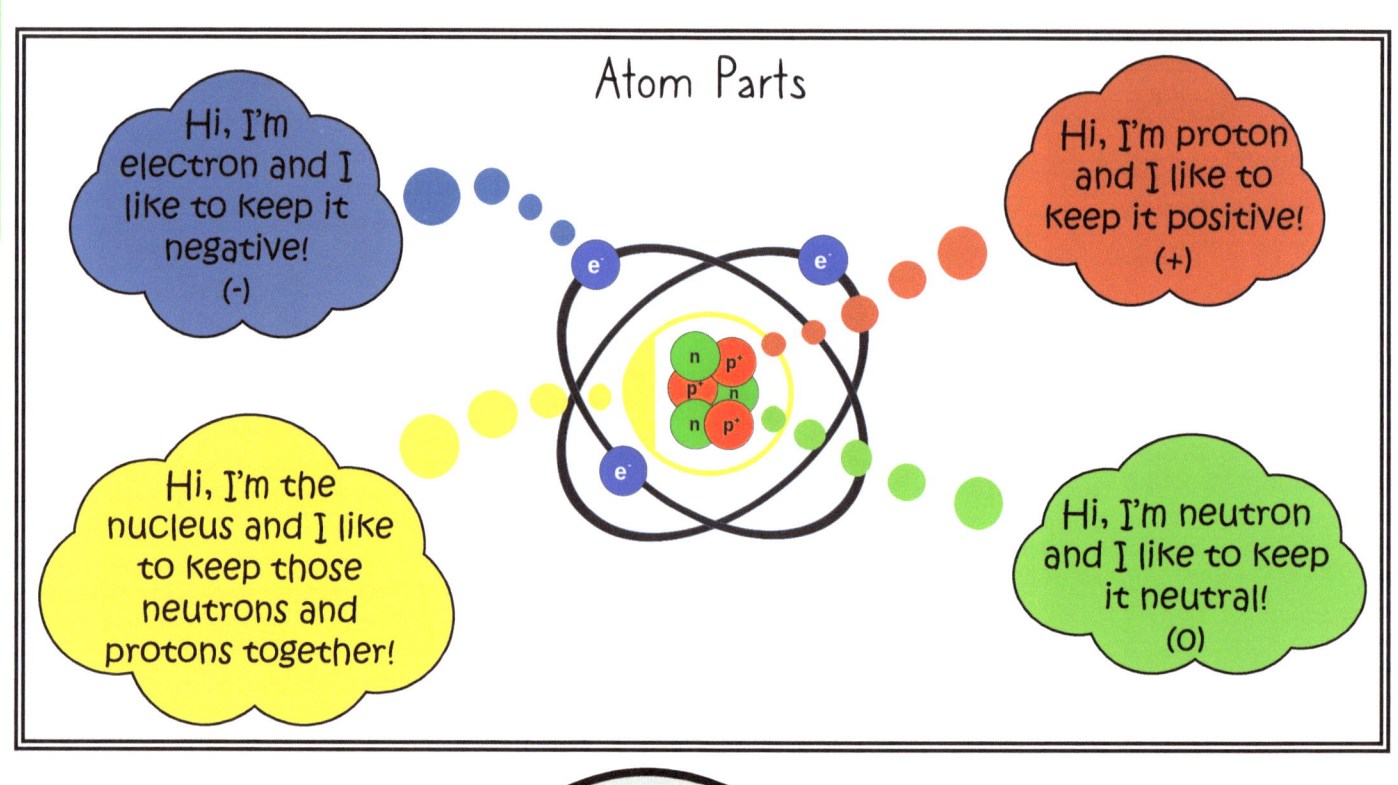

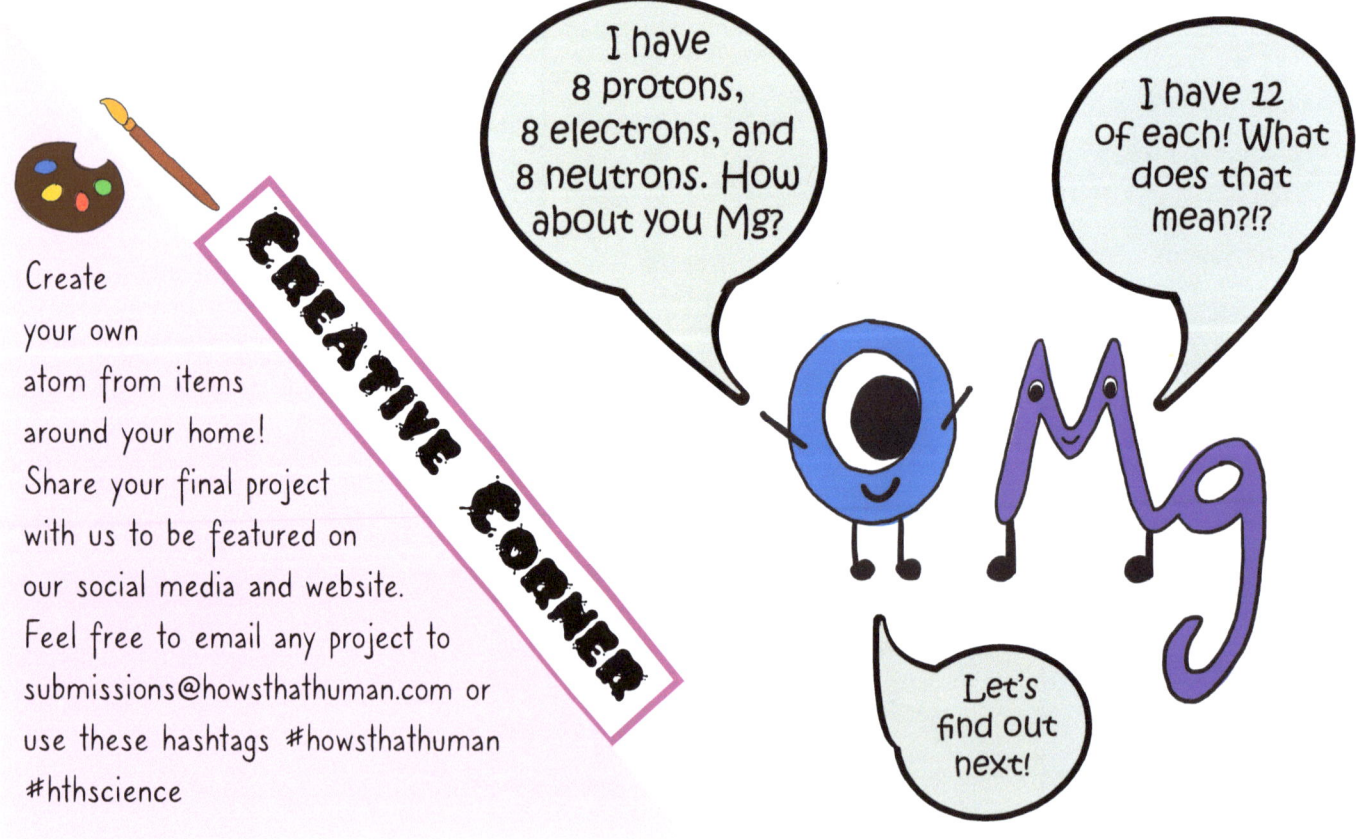

Create your own atom from items around your home! Share your final project with us to be featured on our social media and website. Feel free to email any project to submissions@howsthathuman.com or use these hashtags #howsthathuman #hthscience

SECTION 2: Let's Get Elemental!

 **QUICK QUESTION:** What is an element?

→ An ELEMENT is a pure substance that cannot be chemically broken down. Each element is made up of one type of atom.

 "That means all the oxygen atoms have the same number of protons."

 "Wait, so that means I'm different from you because I have 12 protons?"

"Yep, and I only have 8!"

Scientist Spotlight

Dmitri Mendeleev was a Russian Chemist who came up with a way to organize all the discovered elements based upon physical and chemical similarities. It is called the Periodic Table, pictured here:

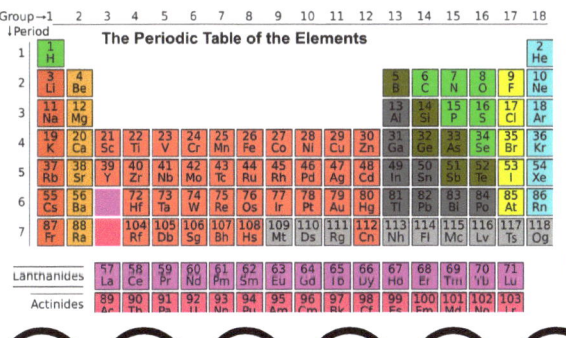

Ready to Research?!

What are the top 10 elements found in the human body?

Search Suggestion: elements in the body

DID YOU KNOW: Only 4 elements make up 96% of our body weight?!

SECTION 3: Properties of Matter

MATTER

↓ Made of atoms

↓ Elements are made of 1 type of atom

↓ Elements can exist by themselves or together with other elements

↓ Have different physical or chemical properties by themselves or together with other elements

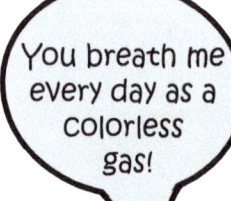

"You breath me every day as a colorless gas!"

"When oxygen is with me, together we are a white powder. But when he is with the hydrogen twins, they are a clear liquid otherwise known as H_2O or water."

Data Dive

Rusting of Nails

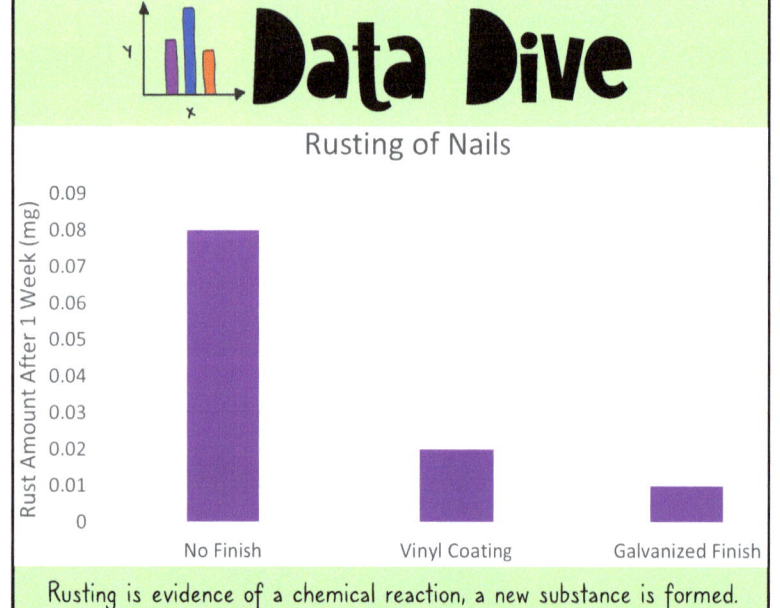

Rusting is evidence of a chemical reaction, a new substance is formed. The element iron reacts with oxygen and water so we have learned to protect nails with a coating or a finish.
What is one conclusion you can come up with by looking at this data?

INTRODUCING THE HYDROGEN TWINS!!

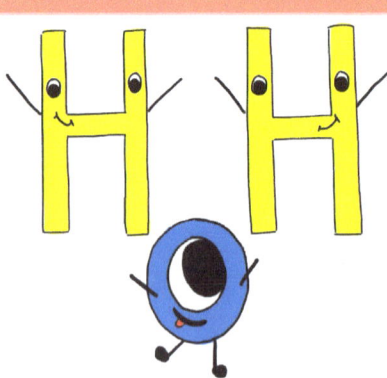

SECTION 4: Hands-On From Home

Lab 1: Can You See It?
Goal: You will use a model to show that matter is made up of particles too small to be seen.

Materials:

Option 1	Option 2
- Baking soda	- Dissolving cold medicine tablet
- Vinegar	- Water

For both: balloons, either a plastic water bottle or test tube, and measuring tools (1/4 cup, tbsp, tsp)

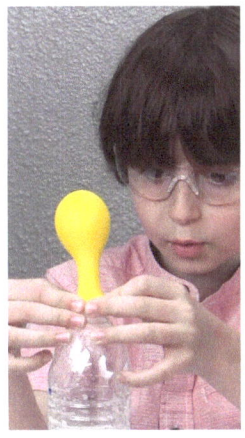

Procedures Option 1:
1. Pour vinegar in your small plastic water bottle (1/4 cup) or test tube half full.
2. Have your balloon ready to place on the mouth of the water bottle or test tube.
3. Pour baking soda into the test tube or water bottle. The amount will vary depending on if you are using the water bottle (2 tbsp) or test tube (1 tsp). Using a funnel makes it easier if you have one or make one from paper.
4. Quickly place balloon over opening and observe what happens.

Procedures Option 2:
1. Pour water in your plastic water bottle or test tube until it is half full.
2. Have your balloon ready to place on the mouth of the water bottle or test tube.
3. Drop the dissolving cold medicine tablet into the water bottle or test tube (will probably have to break it up into pieces).
4. Quickly place balloon over opening and observe what happens.

Lab 2: Physical Properties
Goal: Identify materials based upon physical properties.

Materials:
- 3 different types of apples (or any food where you can get 3 different types)
- Ruler
- Apple cutting utensil (have an adult help)
- Journal and writing utensil

Procedures:
1. Create a data table for the 3 apples with these categories: color, shape, volume, hardness, texture, odor, and taste.
2. First write down the color of each apple, its shape, and how hard it feels.
3. Next using a ruler you will approximate how much space it takes up, or volume.
-You measure the length, width, and height and multiple all 3 numbers together.
4. Now slice a chunk of each apple and record on your data table what it smells like to you.
5. Take a bite and record the taste and texture (is it smooth, rough, etc.)
If you have a kitchen scale, you can also measure weight or mass.

PART 2
COMPOUNDS + BONDING = YOU

When you develop, all your cells are told what to become. For example, when stem cell grows up it could be a....

Heart Cell   Muscle Cell   Brain Cell

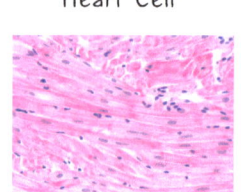 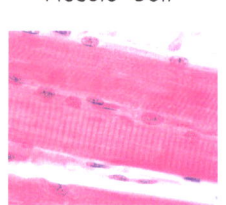 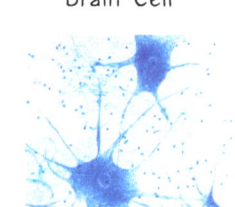

Hi there! I'm a human stem cell. Basically, I'm the building block of your body and I'm made of atoms myself!

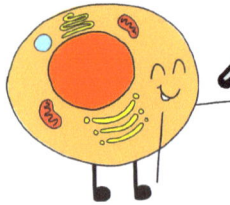

But how does this transformation happen?

Through **CHEMICAL COMMUNICATION** between elements and compounds or molecules!

QUICK QUESTION: What is a compound or molecule?

→ A COMPOUND is two or more atoms of different elements held together with chemical bonds. A MOLECULE is two or more atoms held together by chemical bonds but can be the same element.

I'm not just made up of single elements, but rather a BUNCH of elements stuck together in compounds like H_2O, CO_2, and $C_6H_{12}O_6$.

So what do chemical communication and chemical bonds have in common??
It's time to let electron take the stage!

SECTION 1: Element-To-Element

Atoms share or give electrons in order to make a bond with each other. These bonds hold the atoms close together until a stronger force pulls them apart. There are two common bond types:

1. Covalent - atoms share electrons with each other

2. Ionic - one atom gives its electrons to another

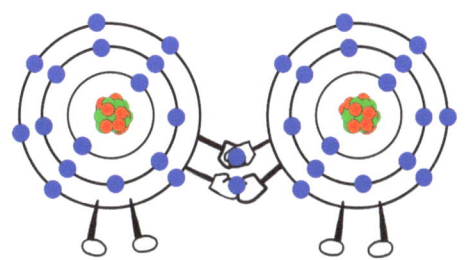

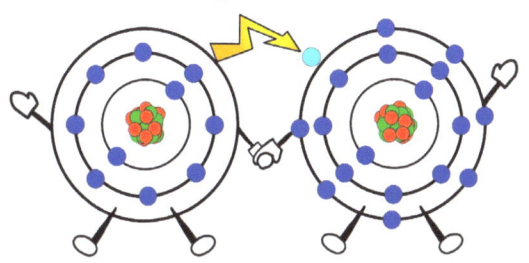

Because electrons carry a negative charge with them, the element who gets the electron turns negative, while the element which gave up the electron becomes positive. They are now called IONS. Let's look at the ions in seawater...

CREATIVE CORNER

Make your own cartoon or comic of chemical bonding. Don't forget to share with the HTH community via email or social media. Email any project to submissions@howsthathuman.com or use these hashtags #howsthathuman #hthscience

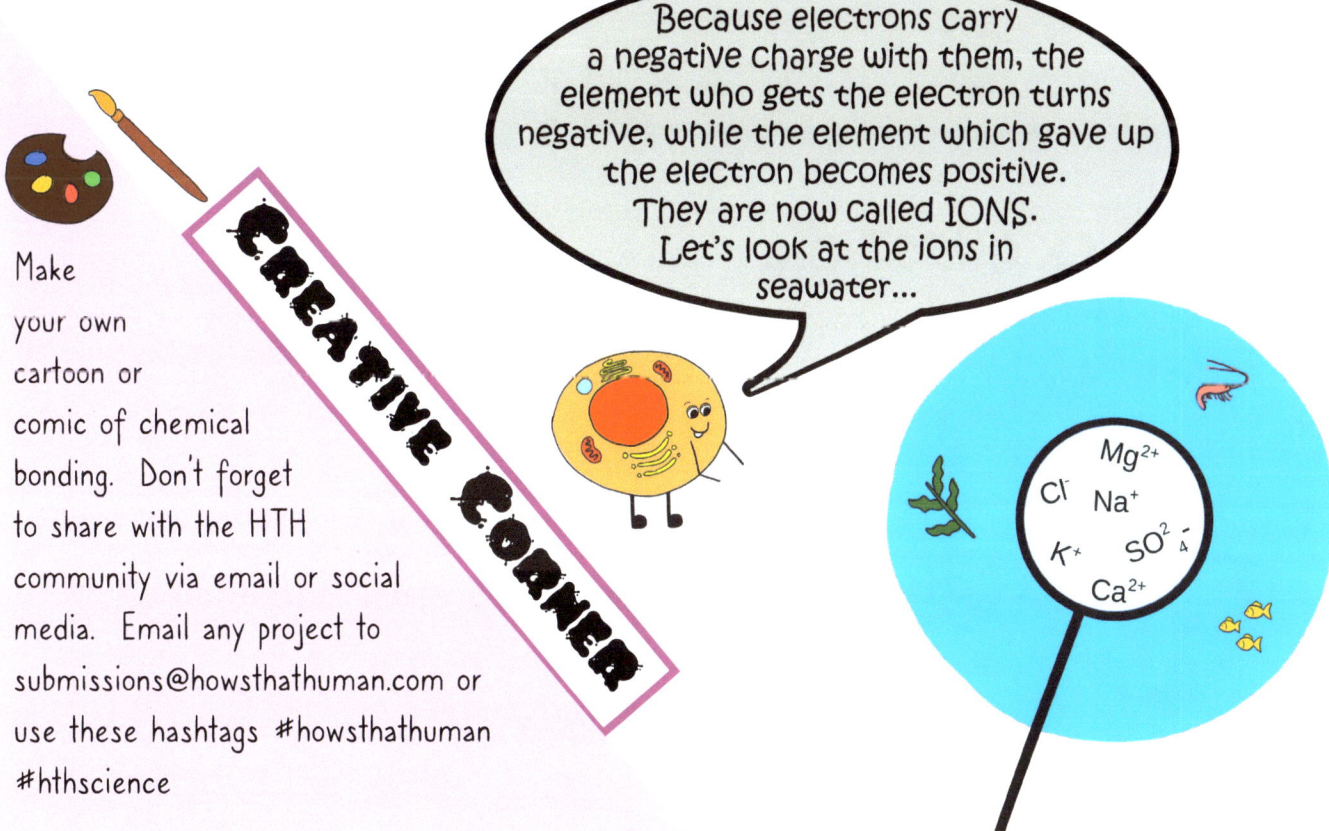

12

SECTION 2: Models

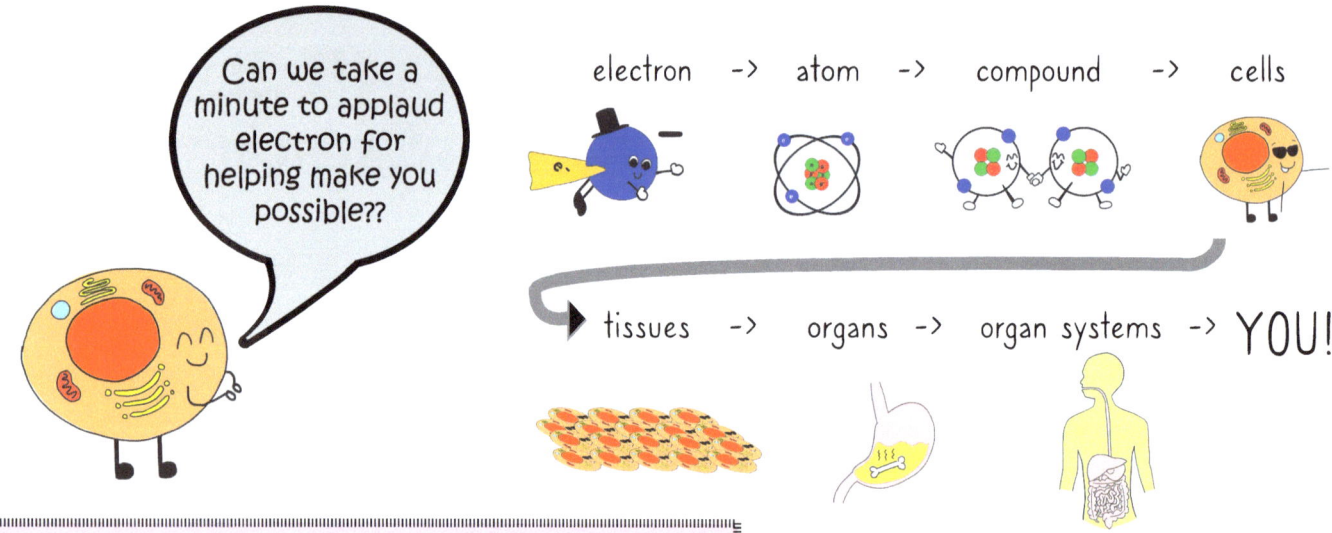

We have 3 different methods to model chemical bonding:

LINES   BALL AND STICK   SPACE FILLING

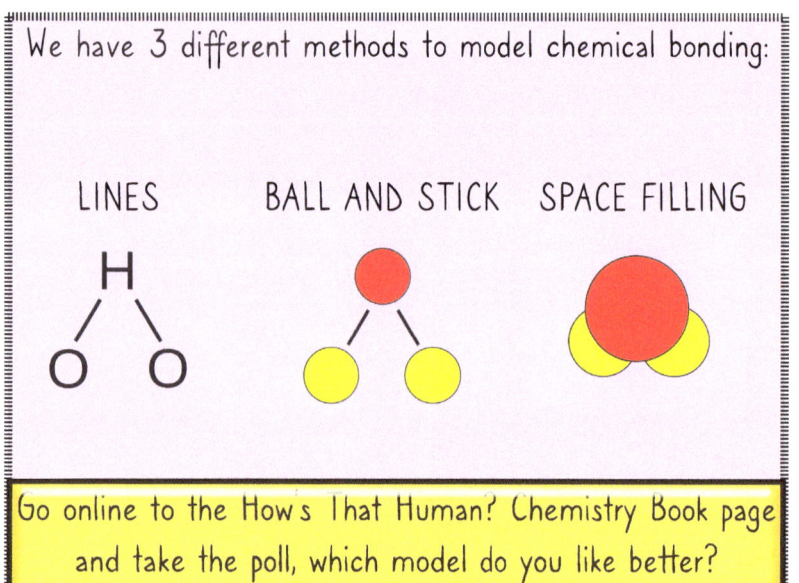

Go online to the How's That Human? Chemistry Book page and take the poll, which model do you like better?

Scientist Spotlight

August Wilhelm von Hofman, a German chemist, was the first to build models of molecules using sticks for bonds and balls for atoms in 1860s.

 **QUICK QUESTION:** In science, not the runway, what are models and why are they used?

-> MODELS represent ideas, objects, or processes that cannot be experienced directly or easily in a lab.

Ready to Research?!

Another type of bond is the metallic bond. How does it relate to electricity?

Search Suggestion: metallic bonds, electrical current

DID YOU KNOW: Hydrogen bonds hold your DNA together and explain many water properties?!

13

SECTION 3: Hands-On From Home

Goal: Practice making bonding models with important biological molecules!

Materials:
- A package of gum drop candy, or any thick gummy candy in multiple colors (could use colored marshmallows)
- Tooth picks
- Modeling Kit Optional

Procedure:
1. Separate your candy into piles by color, you will need 4 different colors.
2. The main elements found in all biological molecules are hydrogen, oxygen, nitrogen, and carbon. Assign them each a color and use the largest number of one color as hydrogen.
3. Start making your structure with the correct elements (gum drops) and bonds (toothpicks) between the elements using the diagrams below as a guide. Here are some hints:
   - H (hydrogen) has only one electron to share, so one toothpick sticking out of it
   - O (oxygen) needs two electrons to be happy, so two toothpicks will stick out of it
   - N (nitrogen) needs three electrons to be happy, so three toothpicks will stick out of it
   - C (carbon) can share four electrons, so four toothpicks will stick out of it

The Backbone of Fats Glycerol

The Amino Acid Glycine (there is a carbon sharing 2 electrons with an oxygen at the same time, this is called a double bond and you need 2 toothpicks)

The Carbohydrate Glucose (makes a ring structure)

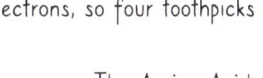

Can you spot the difference?

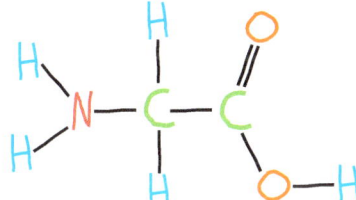

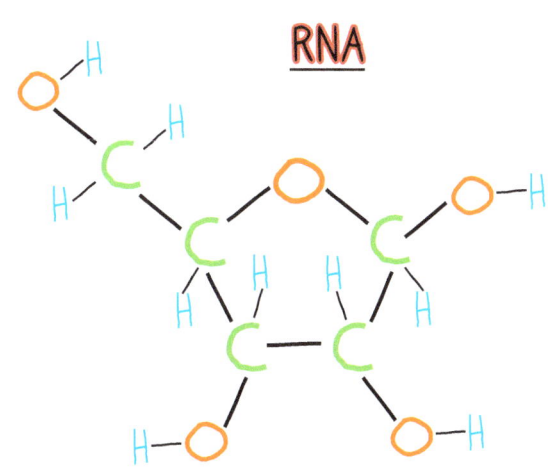

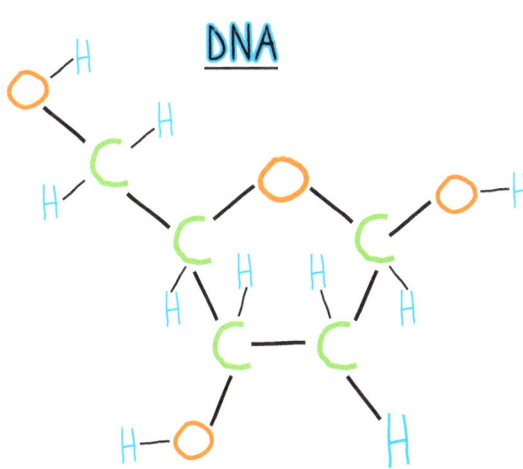

PART 3
States of Matter
and
Chemical Reactions

"Oxygen here and I'm back with the hydrogen twins as the molecule water. We went through a chemical reaction where we share electrons. What type of bond is that again?"

"Covalent!!!"

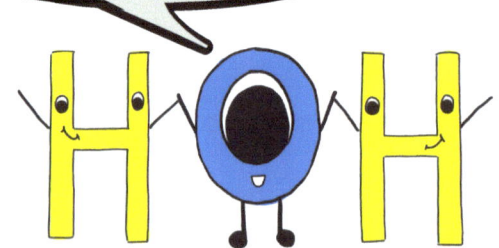

Matter can exist in many forms and can change between those forms. These forms are called the states of matter.

SOLIDS	LIQUIDS	GASES
Molecules are packed together in a fixed shape	Molecules slide past each other and have no particular shape	Molecules are far apart and move freely with a lot of energy

"As water we are essential to life, and happen to be a great example of the three main states of matter!"

  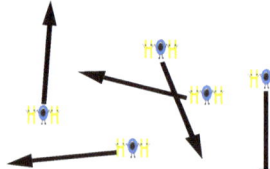 

Ready to Research?!

There is yet another phase of matter, and the most abundant, called plasma. What are 3 ways humans use or interact with this fourth state?

Search Suggestion: plasma state of matter

DID YOU KNOW:
For a diamond to turn into a liquid, you would have to heat it to over 4000C° !!!

15

SECTION 1: Changing Of The States

QUICK QUESTION: How can matter change states?

-> By increasing or decreasing temperature or pressure.

Temperature is the measurement of the amount of energy the molecules have based upon their movement. The faster they move, the more energy, the hotter the temperature!

Pressure is the measurement of how many times molecules bump into the walls of their container. If you make the container smaller, they will run into the walls more increasing the pressure! POP!

Want More? Play with changing states via the How's That Human? Chemistry Book website!

Data Dive

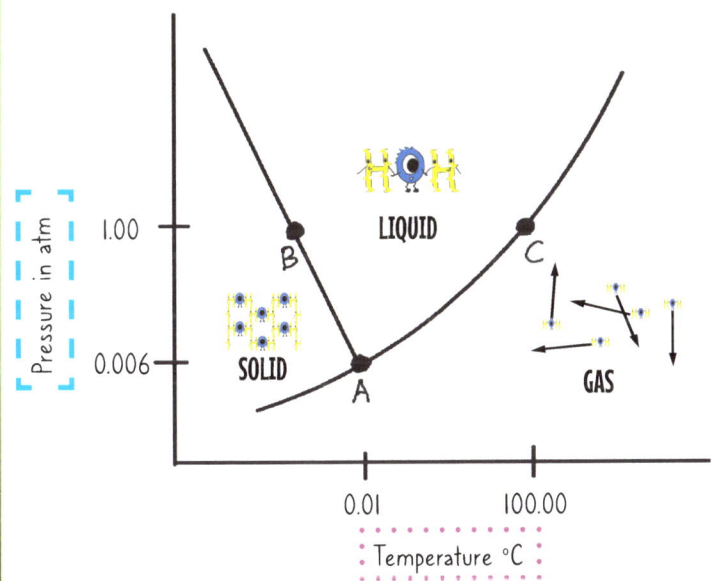

Chemists use phase diagrams to look at the specific temperature and pressure it takes for molecules to go from phase to phase.

This is the phase diagram for water showing what phase water is in at certain temperatures and pressures.

Point A is really cool! It's the temperature and pressure at which water molecules can go between all 3 states of matter and it's called the Triple Point.

Which point between B and C shows the temperature and pressure at which water boils or goes from a liquid to a gas? What temperature does this occur at? What pressure does this occur at?

Scientist Spotlight

In 1924 Indian scientist Satyendra Nath Bose teamed up with Albert Einstein to predict a 5th state of matter. It was named the Bose-Einstein condensate, but wasn't created in a lab until 1995!

SECTION 2: Chemical Reactions

When compounds or elements go through a chemical reaction, new substances are formed. The atoms gain, lose, or switch partners with other atoms all through the sharing or taking of those electrons! Check out this example of the formation of water, and see how the states of matter can change...

REACTANTS (the substances being combined)

The ARROW indicates a chemical reaction is happening.

PRODUCTS (the substances being formed)

The hydrogen twins bond together as a gas.

Oxygen pairs up with another oxygen as a gas.

Together they form a liquid!

There are many ways you can tell a chemical reaction has occurred:

Want More? Watch chemical reaction videos from the HTH Chemistry Book website!

1. LIGHT IS RELEASED

2. A COLOR CHANGE HAPPENS

3. A PRECIPITATE FORMS AND SETTLES ON THE BOTTOM

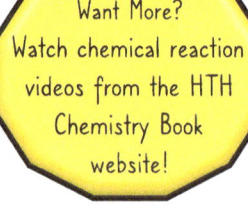

4. BUBBLES FORM

5. TEMPERATURE RISES OR FALLS

 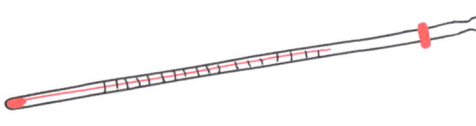

6. IT SMELLS OR TASTES DIFFERENT

CREATIVE CORNER

What does the tongue tell the brain about taste? Create a poster board describing the chemistry of taste. What parts of the tongue taste sweet, salty, bitter, or sour? How does the brain know the difference? Feel free to email any project to submissions@howsthathuman.com

SECTION 3: Hands-On From Home

Lab 1: Can you make a new substance when you mix two or more substances together?

Think about mixing sand and water, is anything new created?

Goal: To perform an experiment that shows you can create something new and something exciting!

Materials:
- water bottle, 16oz is perfect
- 1/2 cup of 20 volume hydrogen peroxide (6% solution you can find in a beauty supply store or hair salon), this can be hard to find so you can use the hydrogen peroxide from the store (the reaction just won't be as big)
- 1 tablespoon of dry yeast (1 packet)
- 1 tablespoon of liquid dish washing soap
- 3 tablespoons of warm water
- food coloring
- small cup
- safety eye wear
- funnel (optional)

Procedures:
1. You must wear the safety eye wear while performing this reaction, so put them on now.
2. Pour the 1/2 cup of hydrogen peroxide into the empty water bottle.
3. Add 8 drops of your favorite food coloring to the bottle.
4. Add 1 tablespoon of liquid dish soap into the bottle and carefully swish it around until it mixes in.
5. In the small cup, put 3 tablespoons of warm water and 1 tablespoon of yeast and mix for about 30 seconds. Try and make it liquid with little clumps, but don't be too rough with the yeast!
6. Put the water bottle onto something you can make a mess on. Place a funnel on the bottle (if you have one) and pour in the yeast mixture. Remove the funnel and watch the fun begin!!

NOTE This is a great experiment for showing that heat is also an example of a chemical reaction occurring. You can run your hands through the "toothpaste" and feel the heat!

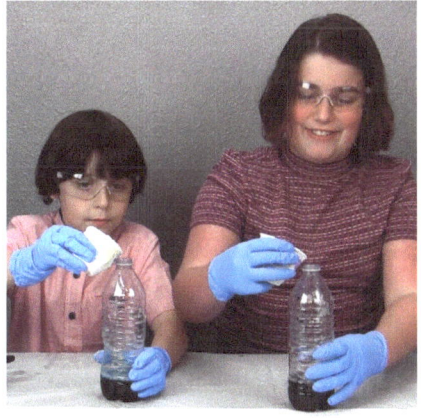 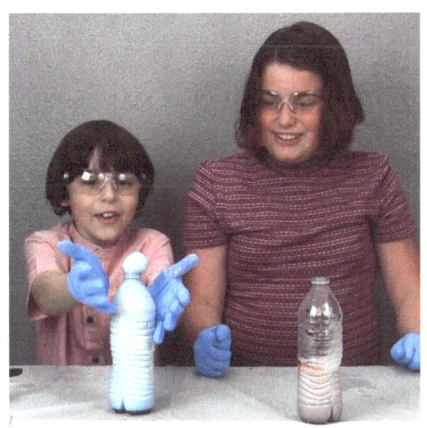 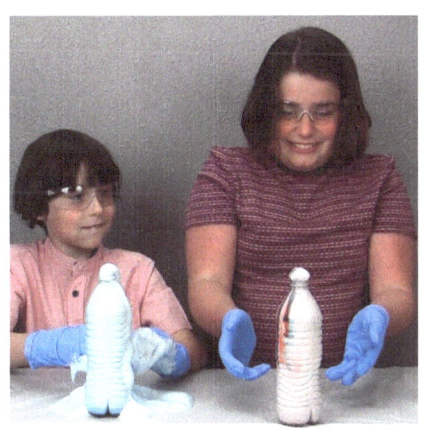

SECTION 3: Hands-On From Home

Lab 2: Through chemical reactions we know that matter can change forms (solid, liquid, gas). For example, our baking soda and vinegar experiment in Part I was a solid mixed with a liquid and a gas formed. Through the reaction, the total amount of matter stayed the same. You collected gas inside the balloon which is still matter, but obviously very light weight.

> This is called the Law of Conservation of Matter!

Think about popcorn, it may change form from a seed to the puffy yumminess we eat, but if you were to weigh the bag before and after popping, it would be the same as long as the bag remains sealed. Test it! However, think about this: what happens if the bag isn't sealed?

Goal: To illustrate the Law of Conservation of Matter with homemade candy!

Materials:
- Jar or large cup
- Source of heat like a stove or microwave
- Pot or cup to heat the water in
- Sugar
- Water
- Measuring cup
- Clean string
- Food Coloring
- Pencil
- Scissors

Procedure:
1. Heat 1 cup of water on the stove or in the microwave to boiling. Be careful and complete with an adult.
2. Mix in 3 cups of sugar until it is dissolved, where did it go?
3. Add a few drops of food coloring.
4. Pour the mixture into the large cup or jar
5. Tie your sting to your pencil and cut it to be as long as 2/3 of the jar. Lay the pencil over the opening so that the string hangs down into the mixture, not touching the bottom. Let it soak for about 30 minutes. Pull it out and let it dry.
6. Let your solution cool to room temperature and place the now hardened string back into the jar with the pencil across the opening.
7. Wait a week and pull it out!

PART 4

MIXTURES and NUTRITION

Hi! I'm Chef's Hat and I'm your guide for this chapter. Sorry if you get hungry!

Last chapter you were mixing things together, but did you know there are different types of mixtures in chemistry??

And that the chemical reactions in your body rely on the pure substances and mixtures you take into your body to function??

⭐❓ **QUICK QUESTION:** What are pure substances?

→ They are elements and compounds that keep their properties like the water you drink, the salt you eat, and the oxygen you breathe.

Visit the HTH Chemistry Book page Community Board and let us know your favorite mixture or pure substance you like to eat or drink!

Ready to Research?!

Our blood is an amazing mixture of many things, what are they?

Search Suggestion: parts of blood, types of blood cells

DID YOU KNOW: Almost 100% of Central and South American populations have O blood type!?

SECTION 1: Mixtures and Solutions

There are two types of mixtures.

Homozygous - look the same throughout!

Heterozygous - can still see original pieces!

QUICK QUESTION: What is a solution?

-> A SOLUTION is two or more substances mixed together that cannot be easily separated. A <u>solute</u> (something being mixed into something else) is combined with a <u>solvent</u> (something that can dissolve another substance).

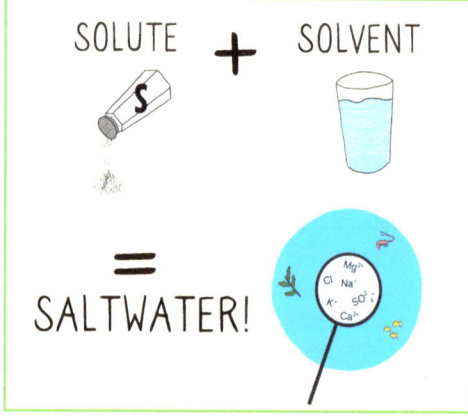

SOLUTE + SOLVENT = SALTWATER!

Chemists use a lot of solutions in their research and need ways to mix or dissolve things faster. Here is how they do it...

CONCENTRATION
You can change the amount of solute you mix in!

MIXING
The rate at which you mix can make a difference!

TEMPERATURE
Heating a solution can speed up dissolving!

CREATIVE CORNER

Ready to show off your scientist skills?! Using the Scientific Method and Poster Guide in the appendix of this book (page 39), design an experiment to test the effects of changing concentration, temperature, OR mixing on the rate of reaction (like how fast something can dissolve). Make sure to share your hard work at submissions@howsthathuman.com to be featured in our Virtual Science Fair and earn a certificate!

SECTION 2: Separation

Think about frying an egg....

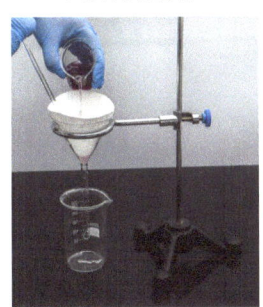

Now that we have mixed things together and watched chemical reactions happen, can we go backward? Here are a few ways chemists try and reverse a chemical reaction or separate a mixture.

FILTRATION

EVAPORATION

DISTILLATION

Hands-On From Home

Goal: To practice separating mixtures with chromatography. Chromatography separates mixtures based on how quickly the individual parts move up a paper with fluid.

Materials:
- Chromatography filter paper or coffee filters
- Colored hard candy or different pens/markers/food coloring
- Water (you can add salt and make a salt water solution and try that or rubbing alcohol too!)
- Toothpicks
- A glass or beaker
- Ruler
- Pencil
- Tape

Procedures:
1. Measure 1 cm from the bottom of the filter paper and draw a line across.
2. Soak your colored candy individually in drops of water to get the dye off (on wax paper works great!)
3. Using the toothpick, dab one of the colored droplets and apply to the line on your filter paper. Do this 3 times to make sure enough dye is on the filter paper. Repeat for your other candies so that each has its own spot on the line.
4. If you are using ink, all it takes is one quick spot on the paper from each maker/pen.
5. In your glass pour enough water to just cover the bottom. DO NOT go higher than the line on your filter paper!
6. Tape your paper to your pencil and place inside your glass or beaker where the solution is touching the paper but below the line with the pencil resting on top to hold it in place.
7. As the water approaches the top of the paper, take the paper out and let it dry. Observe your results!

SECTION 3: Nutrition

Good news! Your body can go backward, breaking down what you eat into the nutrients needed to keep your body going!

These nutrients include all the elements and compounds needed to maintain life processes. So basically, you are what you eat!

Beyond the building blocks of DNA, proteins, and fats, your body needs vitamins and minerals to make sure your body can be the best it can be!

Scientist Spotlight

In 1910, the first vitamin complex was isolated by Japanese scientist Umetaro Suzuki. It's name? Vitamin B1!

Data Dive

COMPOSITION OF THE HUMAN BODY

- Carbohydrate
- Minerals
- Fat
- Protein
- Water

1%, 4%, 10%, 20%, 65%

What percentage of the human body is made up of minerals?

QUICK QUESTION:
What is metabolism?

→ METABOLISM is the term used to describe all the chemical reactions involved in keeping your body working. A lot of times it refers to the amount of energy you release as a result of these chemical reactions, which is of course needed to keep your body running as well.

CREATIVE CORNER

Create an INFOGRAPHIC on a vitamin of your choice. Include sources of food and where it is used in the body.

PART 5
BODY CHEMISTRY

> RECAP!
> 1. You are MATTER, and matter for that matter!
> 2. Matter is made up of ATOMS that share or give electrons between ELEMENTS and COMPOUNDS as BONDS
> 3. With bonding you have many STATES OF MATTER and form new substances through CHEMICAL REACTIONS
> 4. Many PURE SUBSTANCES and MIXTURES provide your body with what it needs to grow and perform all your activities

Time to look even closer at the human body, and how other areas of chemistry are involved!

Hi! I'm parietal cell and I'll be your expert on acids and bases. I'm a specialized cell in your stomach that makes a powerful acid called hydrochloric acid to help break down food. This isn't the only acid in your body however!

Hi! I'm muscle cell and I get to teach you about polymers. I'm made mostly of proteins, and proteins are made out of amino acids. When chains of amino acids are made, its like a polymer. Did you know that you use many different polymers everyday?

Hi! I'm immune system cell and I can't wait to tell you all about heat and temperature in a branch of chemistry called thermochemistry. Did you know that the reactions of metabolism also release heat? Touch your skin!

Hi! I'm nerve cell and we get to talk all about electricity and chemistry! I'm made the way I am to make sure electricity moves safely and quickly throughout your body! Do you know how fast signals pass through me?

SECTION 1: Acids and Bases

QUICK QUESTION: What is an acid?

→ A chemical substance that tastes sour and reacts with bases and metals.

QUICK QUESTION: What is a base?

→ A chemical substance that is slippery to touch, tastes bitter, and reacts with acids to form a water and a salt.

Acids can also harm you! I must be pretty strong to be making it and releasing it in the stomach!

Welcome back hydrogen!!
How do we know if something is an acid or a base?

Well, if there is a lot of me in the fluid it's an acid. If there is few of me, it's a base. We measure my amounts using something called the pH scale.

Scientist Spotlight

Iranian alchemist Jabir ibn Hayyan was one of the first people to experiment with chemistry. He discovered hydrochloric acid (the same acid in your stomach) and was the first to purify gold. We still use some of his tools in the lab today!

Ready to Research?!
When acids and bases react together, its called a neutralization reaction. So how do antacids work?

Search Suggestion: chemistry of antacids, antacids neutralization

DID YOU KNOW: Human blood is slightly basic?!

25

"I'm not the only cell that releases acids in your body. In fact there are many acids and bases inside you performing chemical reactions to keep you alive! Here are a few examples..."

Amino Acids - the building blocks of proteins which make up your body parts and also participate in maintenance of your body.
Fatty Acids - a major part of each of your cells as well as metabolism.
Ascorbic Acid - otherwise known as Vitamin C, it helps your body heal and keeps your immune system strong.

It's important that the amount of acids and bases in your body remains in balance. What happens when that balance is disrupted?

Headache	Confusion
Shortness of Breath	Lack of Appetite
Increased Heart Rate	Fruity-smelling Breath
Muscle Twitching	Numbness
Tingling	Drowsiness
Nausea	Vomiting

"Check out these symptoms!"

 **QUICK QUESTION:** What is homeostasis?

-> HOMEOSTASIS is how your body keeps itself in balance with the right amount of substances so you are healthy. It does this through many different chemical reactions and is controlled by all the chemical communication between your body systems.

Let's say you get too many hydrogens in your blood, becoming more acidic and the pH lowers.

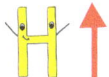

Brain monitors the situation and increases your breathing, good 'ol chemical communication!

Chemical reactions in the blood take out the extra hydrogens and the pH rises returning the balance.

Let's say there aren't enough hydrogens in the blood, the pH rises becoming more basic.

Brain again monitors the situation and slows down your breathing.

Different chemical reactions in the blood add hydrogens back in and the pH drops back to the balance.

Hands-On From Home

We test for acids, bases, and pH values by using indicators, and you can make one at home!

QUICK QUESTION: What is an indicator?

-> An INDICATOR is a compound that changes color when it reacts with an acid or a base. We have optimized tests that use multiple indicators in one it get a more precise measurement, called an Universal Indicator.

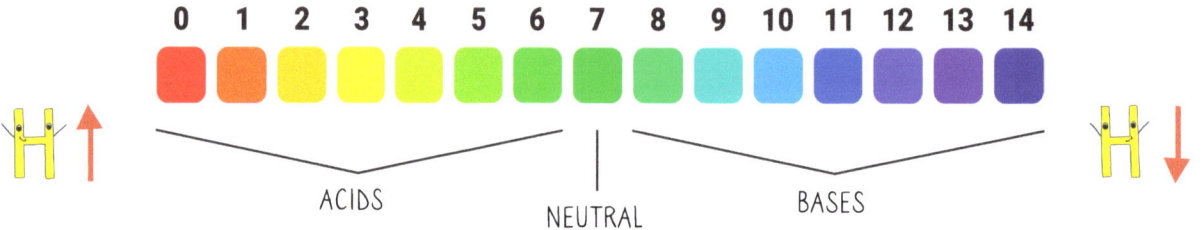

The colors tell us if something is an acid or a base, along with the pH value of how strong or weak that acid or base is. The strongest acid would register as 0, while the strongest base as 14. Acids have a lot of hydrogen ions and bases do not.

Goal: To use a natural indicator to determine whether household fluids are acids or bases.

Materials:
- Red cabbage
- Pot
- Stove
- Water
- Random fluids from around your house including things you drink and things you clean with
- Gather enough glasses to accommodate your fluids from the house, plus one for water

Procedures:
1. Boil your red cabbage in water for about 10 minutes or until the water is dark purple. A half of head of cabbage and 5 cups of water will work great.
2. Remove the cabbage and let the purple water cool down.
3. Pour about an inch of each liquid in its own glass, including one for water.
4. Pour about an inch of red cabbage juice into each cup and observe.
5. Arrange the glasses in a color order that makes sense. With water neutral, neither an acid or base, it should be in the middle. Your acids should all look similar in color and your bases should be a different set of similar colors on either side of your water cup.

SECTION 2: Polymers

QUICK QUESTION: What is a polymer?

→ A substance that has a large number of repeated units bonded together.

> The building blocks of polymers are called monomers. To build a polymer, each single monomer bonds to the next by sharing electrons with each other. To do so they also kick out two hydrogens and an oxygen who bond together as a water molecule. This is a called a condensation reaction.
> Remember the biomolecule amino acid glycine from Part 2? It is one of the building blocks of proteins!

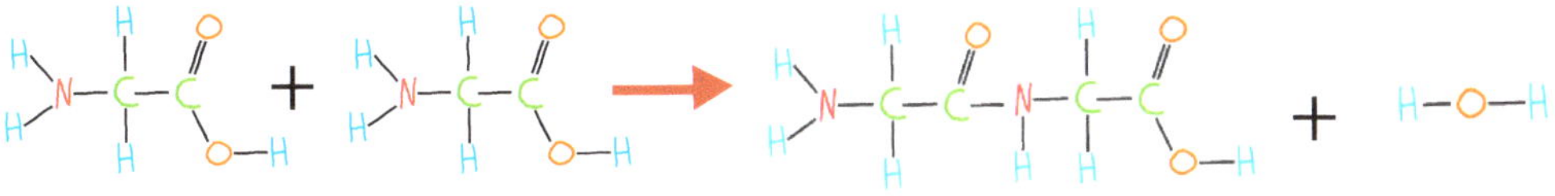

I bet all of these look familiar. They are all made from polymers!

> Condensation typically refers to when water collects as droplets. Is this a good name for how polymers are formed?

CREATIVE CORNER

Make a case! Which of the following is the most important natural polymer? Write an opinion piece, design an Ad Campaign, film a political commercial, etc. Check out the Project-Based Learning Ideas in the appendices on page 37 for more inspiration. Submit your opinion to submissions@howsthathuman.com to be featured on our Great Debate page.

- Collagen - keeps your skin stretchy and smooth
- DNA - the code for life
- Starch - stores the sun's energy that drives life
- Cellulose - the most common compound on earth, keeps plants standing

Scientist Spotlight

Walter Lincoln Hawkins was an American scientist and engineer who helped develop a cable sheath with a polymer that improved the lifespan of telecommunication cables up to 70 years! This allowed telecommunication to move all over the world!

Hands-On From Home

Slime is a great way to show how we can alter the properties of polymers in a fun way! Inside glue the polymers slide easily past each other until you introduce a cross linking agent. The cross linking agent is an ion, remember from Part 2 that means it has a positive or negative charge. As soon as you add a cross linking agent to the glue, charges on the polymers will be attracted to the cross linking agent and movement will be limited.

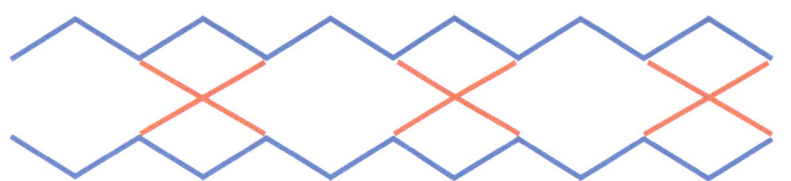

You know what they say about opposites attract!

Goal: To explore the properties of polymers.

Materials:
- Cross linking agent
- 8 oz glue
- Food coloring
- 2 small bowls
- Spoon
- Heat supply
- Water
- 1 tbsp and 1 tsp measuring utensils

Procedures:
1. Warm water and put 1 tbsp of it in your first small bowl
2. Add 1 tsp of the cross linking agent and mix well
3. In the other small bowl pour glue and add food coloring
4. Add in the cross linking agent mixture a spoonful at a time and mix until you get your slime

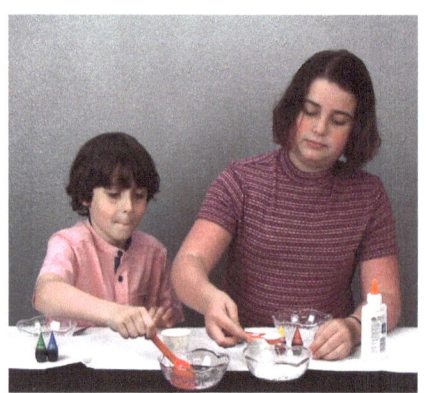

SECTION 3: Thermochemistry

Molecules on earth, like the gases in the air or our skin cells, take in the energy from the sun. You can feel this energy as heat and molecules use this energy to move faster.

Whew! It's hot under the sun, but to be honest, we should be thankful! Inside those waves of light is energy that is used in several ways!

QUICK QUESTION: What is heat?

→ HEAT is the transfer of energy from one substance to the next, from a hotter system to a cooler system.. The more energy, the more movement, the more heat. TEMPERATURE is the measurement of the average amount of movement that molecules have at a given place and time.

Plants take in this energy with carbon dioxide gas and water and go through a chemical reaction called PHOTOSYNTHESIS. The energy is stored in the bonds between elements in the carbohydrate glucose, remember this biomolecule from Part 2? Plants also release oxygen gas in this reaction.

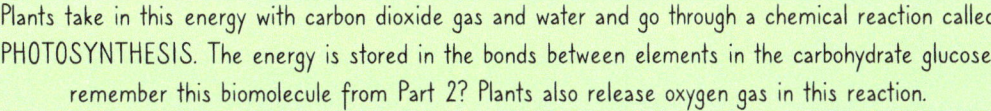

As humans, you eat the plants and release that stored energy by breaking the bonds between the elements. This is a part of metabolism and a chemical reaction called CELLULAR RESPIRATION.

The circle of life depends on those two chemical reactions, and we use that energy to fight the bad guys entering your body!

Ready to Research?!

There is another way the body fights infections, by using heat. What is the purpose of fevers?

Search Suggestion: science of fevers, purpose of fevers

DID YOU KNOW: Bacteria that exist on hydrothermal vents deep in the ocean don't need oxygen to live?!

30

Heat transfers in many ways through the interactions of molecules.

CONDUCTION	CONVECTION	RADIATION
heat passes through physical touch	heat passes through liquids or gases	heat passes through electromagnetic waves

Hypothermia is a condition where your body temperature falls too low. In what ways could you use heat transfer to help someone? Join our online Discussion Board on the Hows That Human? Chemistry website and tell us how you would use either conduction, convection, or radiation to help heat someone back up.

"How many times have you used different amounts of heat to cook food?"

Chemists use heat to provide more movement for more collisions to happen in a chemical reaction. If a reaction is cold, molecules move less and are less likely to run into each other slowing down the reaction.

Scientist Spotlight

Nicolas Léonard Sadi Carnot was a French scientist who changed the world with the invention of the heat engine. His work led to the branch of science called thermodynamics, the topic of this section!

Data Dive

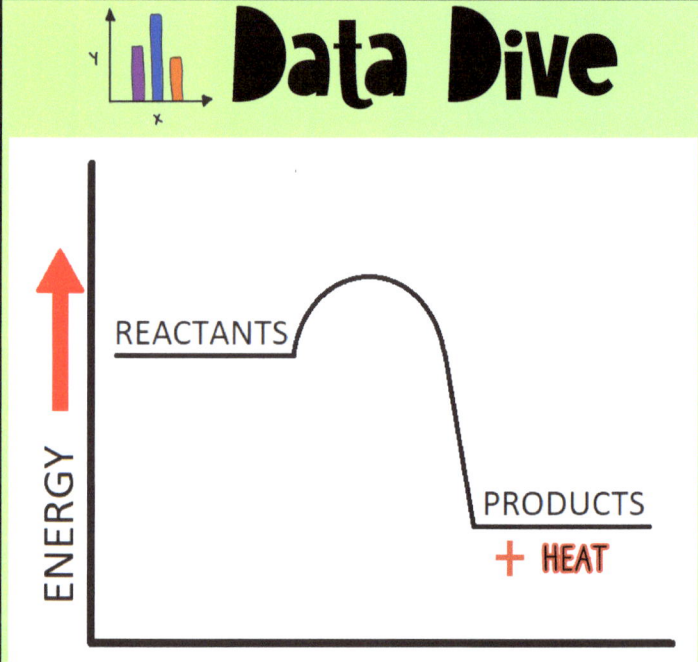

Energy is conserved through chemical reactions just like matter. We can represent reactions that take in or release heat by measuring that energy. If the reactants start with more energy than the products, the reaction is called exothermic because some of that energy is released as heat to the environment with the products. If the reactants start with less energy compared to the products, its called an endothermic reaction because heat was needed to help form the products.
What type of reaction is pictured above? Endothermic or Exothermic?

Hands-On From Home

Lab 1
Goal: To explore how temperature affects the rate of reaction.

Materials:
- 3 glow sticks, same size and color
- 3 glasses
- Ice
- Source of heat and pot
- Water

Procedures:
1. Place water and ice in your first glass
2. Place room temperature water in the second glass
3. Heat water to almost boiling and put in third glass (have an adult help)
4. Snap all 3 glow sticks at once and place as quickly as you can into each of the three water glasses
5. Observe the brightness of reaction by turning off your lights

Lab 2
Goal: To use homemade ice cream to show how temperature changes can affect states of matter.

Materials:
- 1/2 cup milk
- 1/2 cup whipping cream (heavy cream)
- 1/4 cup sugar
- 1/4 teaspoon vanilla or vanilla flavoring (vanillin) or chocolate syrup or fruit
- 1/2 to 3/4 cup sodium chloride (NaCl) as table salt or rock salt
- 2 cups ice
- 1 quart sealable plastic bag (like a zipper-top baggie)
- 1 gallon sealable plastic bag
- gloves
- measuring cups and spoons
- cups and spoons for eating your treat!
- *Optional Thermometer*

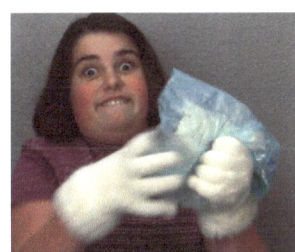
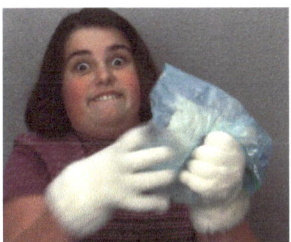
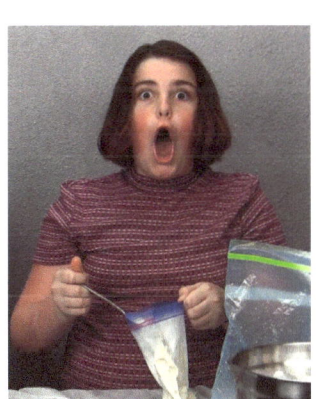

Procedures:
1. Add 1/4 cup sugar, 1/2 cup milk, 1/2 cup whipping cream, and 1/4 teaspoon vanilla to the quart sealable plastic bag. Seal securely!
2. Put 2 cups of ice into the gallon sealable plastic bag. If you have a thermometer, record the temperature of the ice in the gallon bag.
3. Add 1/2 to 3/4 cup salt (sodium chloride) to the bag of ice.
4. Place the sealed quart bag inside the gallon bag of ice and salt. Seal the gallon bag securely.
5. Put on your gloves or use a cloth and gently rock the gallon bag from side to side. The cold will damage your hands, protect yourself!
6. Continue to rock the bag for 10-15 minutes or until the contents of the quart bag have solidified into ice cream. Again, if you have a thermometer, record the temperature of the ice/salt mixture.
7. Remove the quart bag, open it, serve the contents into cups with spoons and enjoy!

SECTION 4: Electrochemistry

"As you saw in the previous section, energy is pretty important and I'm the lucky cell that transports it in your body!"

QUICK QUESTION: What is electricity?

→ Electricity is the flow of electrons or the energy you get as the electrons move from place to place.

How do we get electricity? It starts with...

NONRENEWABLE

Coal

Natural Gas

RENEWABLE

Wind

Water

Oil

Nuclear   Solar

....and is transferred into usable forms for us via POWER PLANTS and BATTERIES

Data Dive

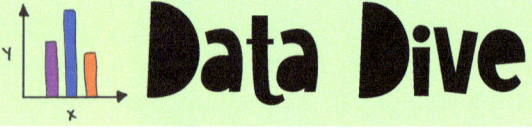

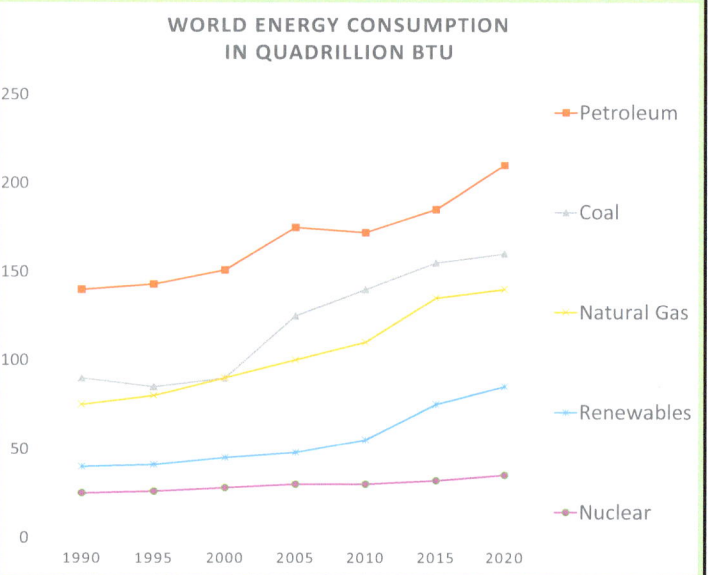

WORLD ENERGY CONSUMPTION IN QUADRILLION BTU

- Petroleum
- Coal
- Natural Gas
- Renewables
- Nuclear

What trends do you see in the data over time? How does worldwide use of nuclear energy compare to coal? Which type of energy is climbing faster more recently? Why are there differences in the data?

Scientist Spotlight

Hong Kong chemist Vivian Wing-Wah Yam's research on organic light emitting diodes has improved mobile phone and computer displays!

33

So how do batteries work? Batteries use chemical reactions to cause one part of the battery to become positive (cathode) and the other to be negative (anode). When you allow a path for electrons to move (circuit) the build of electrons on the negative side move towards the positive side creating an electrical current.

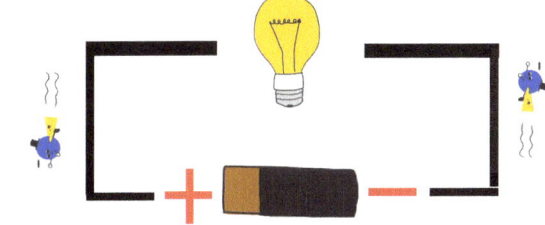

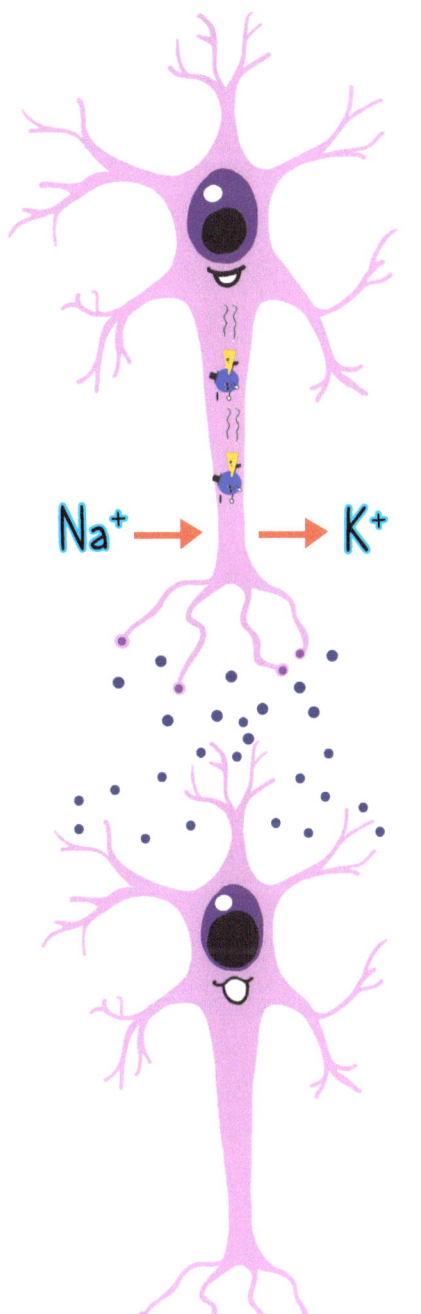

How does this relate to you? Here's where a couple of elements come into play. Potassium K^+ (thanks bananas!) and sodium Na^+ (hello salt!) are used as positive ions and moved across the cell membrane.

This creates an electrical difference that helps electrons move down the cell to the next cell as an electrical current or impulse.

At the end of the cell the electrons energize little chemical messengers called neurotransmitters who cross the synapse, the gap between cells, exciting the next nerve cell.

CHEMICAL COMMUNICATION

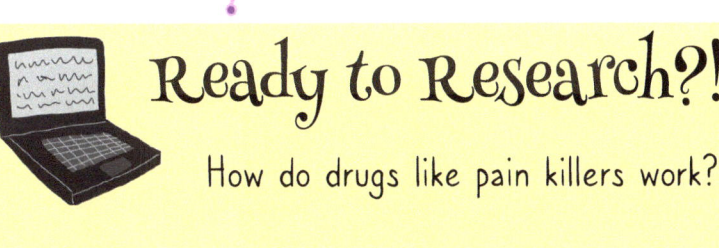

Ready to Research?!

How do drugs like pain killers work?

Search Suggestion: how do drugs affect the brain, how do pain killers work

CREATIVE CORNER

Film a stop motion movie of a nerve impulse. Send it over to submissions@howsthathuman.com to be featured on our Movie Theater page.

Hands-On From Home

Goals: To investigate the properties of electricity through chemistry.

Materials Option 1:
- 10 pennies and 10 nickels
- Paper towels or coffee filters
- 1/4 cup white vinegar
- 1 tbsp salt
- Small bowl
- Scissors
- FOR BOTH: LED pin light small, alligator clips (10 pack)
- *Optional Multimeter* can be used to test for current or voltage

Procedures Option 1:
1. Mix 1/4 cup of vinegar with 1 tbsp of salt in the small bowl.
2. Make sure your pennies and nickels have no dirt or grime on them.
3. Cut tiny squares the size of a penny out of the paper towels and put into the vinegar salt solution.
4. Place a penny down, then a square of vinegar-soaked paper towel on top. It should not be dripping though just thoroughly wet.
5. Place a nickel on top of the paper towel and congrats, you have your first battery! Repeat until you have several layers (penny-paper towel-nickel).
6. Attach an alligator clip to one wire of the LED light bulb, attach a second clip to the other wire.
7. Touch the remaining alligator clip noses to the penny on the bottom of your stack and the nickel on the top of your stack and your light should shine bright!

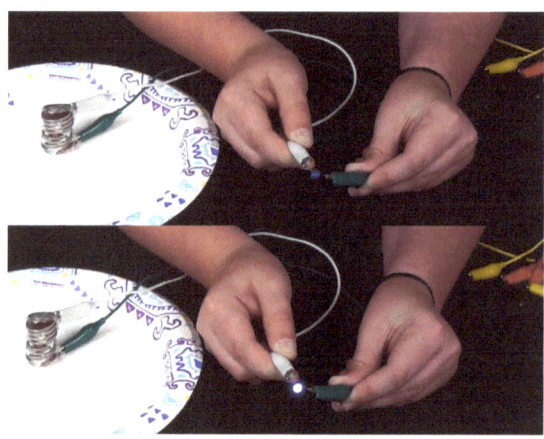

Materials Option 2:
- 4 Lemons
- Zinc/galvanized nails (2in long)
- Copper wire (2in strips) or pennies would work
- FOR BOTH: LED pin light small, alligator clips (10 pack)

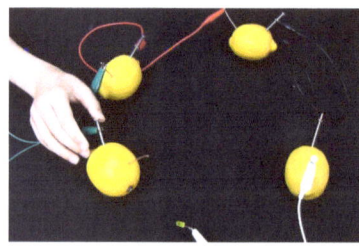

Procedures Option 2:
1. Roll the lemons to make them nice and juicy.
2. Shove a copper wire (or penny) into one side of each lemon and a zinc nail in the other side.
3. Clip one alligator clip to the first lemon copper wire and attach the other end to the next lemon's zinc nail.
4. Repeat until you almost have a full circle with the copper wire of one lemon open and the zinc nail of the last lemon open.
5. Attach one alligator clip to the zinc nail of the last lemon and attach the other side to one of the wires of the LED light bulb.
6. Take another alligator clip and attach it to the copper wire of the first lemon and then the other end to the remaining wire of the LED light bulb, let there be light!

APPENDICES

Category Definitions and PBL Ideas

 QUICK QUESTION These questions are the main idea or key point of the section.

 Data Dive Practice math skills by analyzing data.

 Ready to Research?! Research prompts apply the information for further understanding of the topic.

 Scientist Spotlight Appreciate scientific contributions from around the world.

 Hands-On From Home Step-by-step experimental procedures, materials lists, and pictures.

 Creative Corner This category focuses on Project-Based Learning (PBL) projects.

 Check out below for additional ideas!

Brochure or Pamphlet	Poster	Start a Blog
Map	Presentation	Flipbook
Model	Diorama	Create a Cereal and Cereal Box
Treasure Hunt	Sculptures	Make a Time Capsule
Scavenger Hunt	Poem or Song or Rap	Advice Column
Menu	Timeline	Documentary
Advertisement	Write and perform a skit	Choose Your Own Adventure
Movie/Book/TV Show Review	TV or Radio Commercial	Top 10 List
PSA	Collage or Mobile	Museum Exhibit
News Report	Flowcharts or Diagrams	Paper Chain
Animation	Scrapbook or Photo Album	Album Cover
Stop Motion Animation	Instructional Video	Autobiography/Biography
Comic or Cartoon	Bookmarks	Newspaper or Magazine
Portfolio	Calendar	Invitation or Greeting Card
Awards	Casting Call	Sales Pitch
Banners	Episode of a Reality Show or Game Show	Book Club
Blueprints	Expert Panel Discussion	Puppet Show
App or Video Game Design	Children's Book	Debate
Board or Card Game	Fable or Myth	Mock Court Case
Design a Building or Garden	Help Wanted Poster or Ad	Create a Cheer
Cooking and Baking	Text Message Dialogue Box	Coat of Arms
Exercise	Series of Tweets	Flags
Plan A Lesson	Create a Social Media Page	Hieroglyphics
ID/Merit Badges	Murals	Tests/Worksheets
Illustrated Quotes	Pen-pals	Yearbook
Inventions	Stamps	Questionnaires

 Key Vocabulary

Acid - a chemical substance that tastes sour and reacts with bases and metals
Atom - the building block of matter with a nucleus, protons, neutrons, and electrons
Base - a chemical substance that is slippery to touch, tastes bitter, and reacts with acids to form a water and a salt
Cell Respiration - the process by which some organisms like humans release the energy stored in food for metabolism
Chemical Properties - the result of a chemical reaction changing the compound.
Chemical Reactions - one or more substances are converted into one or more different substances
Chemistry - the study of matter: what it is made of, its properties, and how it changes
Compound - two or more atoms of different elements held together with chemical bonds
Covalent Bonds - atoms share electrons with each other
Electricity - the flow of electrons or the energy you get as the electrons move from place to place
Element - a pure substance that is made up of one type of atom and cannot be chemically broken down
Gases - state of matter where molecules are far apart and move freely with a lot of energy
Heat - the transfer of energy from one substance to the next through conduction, convection, or radiation
Heterozygous Mixtures - can still see original pieces or parts
Homeostasis - how your body keeps itself in balance with the right amount of substances
Homozygous Mixtures - look the same throughout
Indicator - a compound that changes color when it reacts with an acid or a base
Ionic Bonds - one atom gives its electrons to another
Law of Conservation of Matter - matter cannot be created or destroyed, it changes form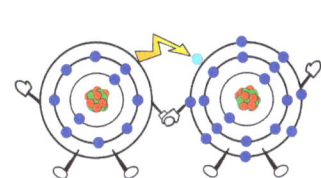
Liquids - state of matter where molecules slide past each other and have no particular shape
Matter - anything that takes up space and has mass
Metabolism - all the chemical reactions involved in keeping your body working
Models - represent ideas, objects, or processes that cannot be experienced directly or easily in a lab
Molecule - two or more atoms held together by chemical bonds and can be the same element
Periodic Table - arranges elements according to their properties
Photosynthesis - the process by which some organisms like plants convert sunlight into food from carbon dioxide and water
Physical Properties - things we can measure or use our senses to describe
Polymer - a substance that has a large number of repeated units (monomers) bonded together
Pressure - the measurement of how many times molecules bump into the walls of their container
Products - the substances being formed as the result of a chemical reaction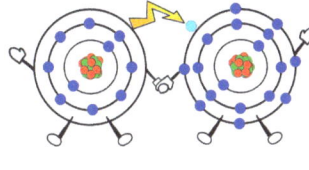
Pure Substances - elements and compounds that keep their properties
Reactants - the substances being combined in a chemical reaction
Solids - state of matter where molecules are packed together in a fixed shape
Solute - something being mixed into something else
Solution - two or more substances mixed together that cannot be easily separated
Solvent - something that can dissolve another substance
Temperature - the measurement of the average amount of energy the molecules have based upon their movement
Volume - the amount of space something takes up

Scientific Method and Poster Design

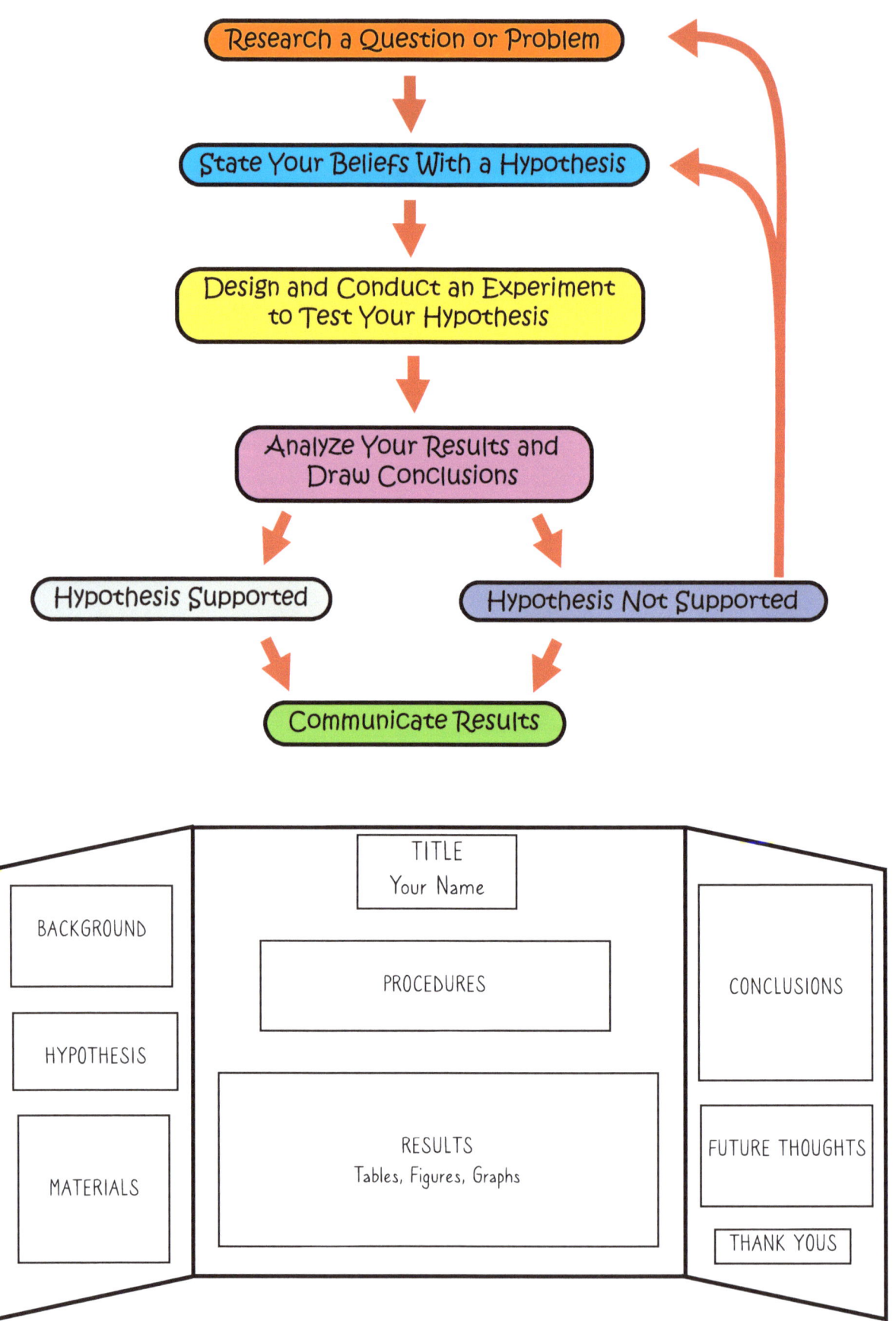

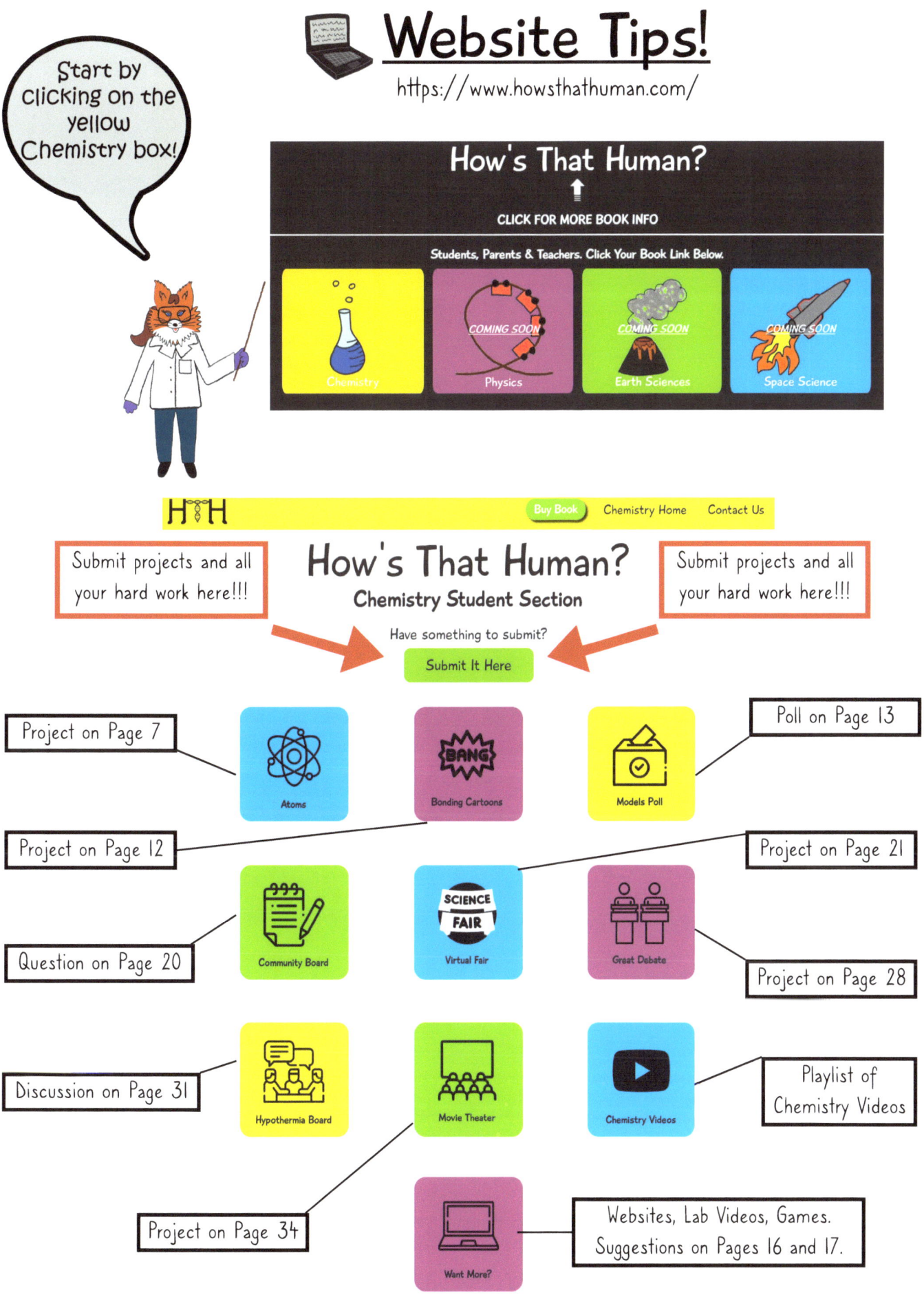

Scientist Spotlight

Author Rita Claire is a former molecular virologist and geneticist who worked on plant virus research involving cereal crops and human genetics projects. She misses research, but has thoroughly enjoyed teaching science K-12 and beyond. Ms.C, as she's affectionately called by her students, is passionate about changing the way we educate and has always dreamed of being an author.
As a recent homeschool mom, she wanted to create a more relatable educational tool for other homeschool or distance learning families, as well as elementary school districts.
Welcome to How's That Human?

Illustrator AM Conroy has a PhD in biology from UCLA and has worked in marine biology, animal physiology, and biomechanics. Her love for animals inspired her to study the movements of amazing creatures such as swimming pufferfishes, running tigers, and hopping kangaroos. Her mom, an illustrator as well, encouraged her interest in drawing as a child, with favorite subjects being snails, fishes, and unicorns. She enjoys illustrating for How's That Human? and hopes that her drawings help excite learners of all ages!

Acknowledgments

Contents Page: August de Richelieu - Pexels
Periodic Table: ExplorersInternational - Pixabay
Heart and Skeletal Muscle Cells: JosLuis - stock.adobe.com
Nerve Cells: sinhyu - stock.adobe.com
pH Scale: dmutrojarmolinua - stock.adobe.com
Filtration: mehmet - stock.adobe.com
Evaporation: LIGHTFIELD STUDIOS - stock.adobe.com
Distillation: totojang1977 - stock.adobe.com